BIRDS
OF PR

Consultant: George E. Watson
Illustrators: Robert Cremins, Tim Phelps

Copyright © 1998 by the National Geographic Society

Published by
The National Geographic Society
John M. Fahey, Jr., President and Chief Executive Officer
Gilbert M. Grosvenor, Chairman of the Board
Nina D. Hoffman, Senior Vice President
William R. Gray, Vice President and Director, Book Division

Staff for this Book
Barbara Lalicki, Director of Children's Publishing
Barbara Brownell, Senior Editor and Project Manager
Marianne R. Koszorus, Senior Art Director and Project Manager
Toni Eugene, Editor
Alexandra Littlehales, Art Director
Amy Donovan, Writer-Researcher
Susan V. Kelly, Illustrations Editor
Jennifer Emmett, Assistant Editor
Mark A. Caraluzzi, Director of Direct Response Marketing
Vincent P. Ryan, Manufacturing Manager
Lewis R. Bassford, Production Project Manager

Visit our Web site: www.nationalgeographic.com

Library of Congress Catalog Card Number: 97-76353
ISBN: 0-7922-3454-5

Color separations by Quad Graphics, Martinsburg, West Virginia
Printed in Mexico by R. R. Donnelley & Sons Company

BIRDS
OF PREY

AMY DONOVAN

All photographs supplied by Animals Animals/Earth Scenes

NATIONAL
GEOGRAPHIC
SOCIETY

INTRODUCTION

You probably see birds in the sky and in yards every day. This book describes a particular kind of bird—the bird of prey. Unlike the finches and chickadees you may find near a feeder, birds of prey are meat eaters. They have special adaptations that enable them to hunt, kill, or scavenge (SKA-venj) other animals. Among these adaptations are their relatively large size, powerful flight, fierce appearance, keen eyesight, hooked beaks, and long, curved claws, or talons (TAL-unz). Birds of prey are also known as raptors (RAP-torz).

In the past, populations of these birds were threatened. To protect livestock, people once hunted raptors. Chemicals used to control insects also harmed them. Today we realize the important role birds of prey play in the environment, and all are protected by law.

Keep binoculars handy and look for roosting owls, soaring hawks or vultures (VUL-churz), and fast-flying falcons (FAL-kunz). Remember that even experts sometimes have difficulty telling some birds of prey apart.

HOW TO USE THIS BOOK

This guide covers raptors in North America; some of them live in other places, too. The book is organized by type of bird of prey. First come vultures, scavengers that eat dead animals. Next come eagles, hawks, and falcons. These birds hunt live prey during the day. A section on owls, most of which hunt at night, begins on page 52. Each spread helps you identify one kind of bird of prey. It tells you about the bird's size, color, and behavior. Some birds of prey stay in the same area all year, but most migrate, or move with the seasons. A shaded map of North America shows each bird's range, and the "Field Notes" entry offers an additional fact about it. If you come across a word you do not know, look it up in the Glossary on page 76.

CALIFORNIA CONDOR

 The California condor once lived in many areas of the continent. As people took over its habitat, this vulture began to die out. Now, it is one of the world's rarest birds.

WHERE TO FIND:
About 100 condors live in zoos; another 40 soar freely over parts of California and Arizona.

WHAT TO LOOK FOR:

✳ **SIZE**
The California condor is about 50 inches long; it has a 9-foot wingspan.

✳ **COLOR**
It has a black body with white under its wings.

✳ **BEHAVIOR**
A female condor lays just one egg every two years.

✳ **MORE**
A condor may eat two or three pounds of dead meat a day.

0000000000000

FIELD NOTES

The condor's wingspan, longest of any North American land bird, helps make it a good soarer.

Adult California condors have orange heads. Like other vultures, the birds have bare heads and necks.

7

TURKEY VULTURE

 Master gliders and soarers, turkey vultures are often seen flying in groups. They are called turkey vultures because their bare red heads and dark bodies resemble those of turkeys.

WHERE TO FIND:
Turkey vultures glide over open country and farmlands from southern Canada through Central America.

WHAT TO LOOK FOR:

✶ **SIZE**
The turkey vulture is about 27 inches long; its wingspan is about 6 feet.

✶ **COLOR**
It is brownish black with gray or silver under its wings.

✶ **BEHAVIOR**
Turkey vultures nest in caves, on cliffs, and in hollow logs on the ground.

✶ **MORE**
Unlike other vultures, turkey vultures can find food using their sense of smell.

Turkey vultures hold their wings upward in a shallow V and seldom flap them in flight.

9

BLACK VULTURE

 Black vultures live in open country and near human settlements, where they scavenge for food in garbage dumps. The tails of black vultures are stubby and square.

WHERE TO FIND:

The black vulture ranges from the eastern United States through Mexico and Central America.

WHAT TO LOOK FOR:

✳ SIZE
Black vultures are about 2 feet long and have wingspans of about 5 feet.

✳ COLOR
They are coal black with white patches under the wing tips.

✳ BEHAVIOR
They are very aggressive and may take over food from turkey vultures.

✳ MORE
Like other birds of prey, black vultures have excellent eyesight.

Black vultures often gather in large flocks and roost together in trees at night.

The bare head of a vulture is easy to clean—a good thing since the food it eats is messy.

GOLDEN EAGLE

 The golden eagle lives in much of North America as well as in Europe, Asia, and northern Africa. Indians of the western plains once used golden eagle feathers in their headdresses.

WHERE TO FIND:
Golden eagles are often found in mountainous regions, particularly in the west.

WHAT TO LOOK FOR:

✴ SIZE
Golden eagles are 30 to 40 inches long with a wingspan of about 7 feet.

✴ COLOR
They are dark brown with gold on their crowns and necks.

✴ BEHAVIOR
They prey mostly on mammals but also kill other birds and eat carrion.

✴ MORE
They mature slowly and do not breed until they are at least five years old.

A dark bill marks
the golden eagle.
It is sometimes
called the king
of birds.

13

BALD EAGLE

The bald eagle is the national symbol of the United States. Endangered until recently, the bird has made a strong comeback. Today you can find bald eagles in many places throughout North America.

FIELD NOTES

Bald eagles build huge nests called aeries (AIR-eez) that can measure as much as 12 feet across.

About four years pass before this eagle develops the white head and tail feathers and yellow bill of an adult.

WHERE TO FIND:
The bald eagle is found in Canada and the United States, usually near seacoasts, bays, and rivers.

WHAT TO LOOK FOR:

✳ SIZE
The bald eagle is about 35 inches long; its wingspan is 7 to 8 feet.

✳ COLOR
Adults have a dark body and a white head and tail.

✳ BEHAVIOR
Bald eagles eat mainly fish, swooping down to snatch them from the water.

✳ MORE
Bald eagles wade in shallow water to grab salmon as they swim past.

SNAIL KITE

The snail kite feeds only on a certain kind of freshwater snail. Kites are small, graceful hawks. They are such good fliers that they gave their name to a popular toy— the kite!

WHERE TO FIND:
Snail kites live near marshes in southern regions of North America. Some live as far north as Florida.

WHAT TO LOOK FOR:

✳ **SIZE**
The snail kite is about 18 inches long; its wingspan is almost 3½ feet.

✳ **COLOR**
Males are blackish gray with red legs; females and young are brownish.

✳ **BEHAVIOR**
The birds use their long, hooked bills to extract snails from their shells.

✳ **MORE**
A snail kite has a white band at the base and tip of its tail.

A young snail
kite prepares to
pull a snail from
its shell. It eats
the animal whole.

NORTHERN HARRIER

The northern harrier is a hawk of the marshes, where it flies low over the grasses in search of mammals, other birds, reptiles, amphibians, and insects. Harriers nest on the ground.

The northern harrier's dish-shaped face directs sounds to the bird's ears.

FIELD NOTES
A ring of feathers around the bird's face helps capture sounds as faint as the rustling of mice in grasses.

WHERE TO FIND:
Northern harriers live near fields and marshes throughout most of North America.

WHAT TO LOOK FOR:

✳ SIZE
Northern harriers are about 20 inches long and have wingspans of up to 4 feet.

✳ COLOR
Females and young harriers are brown; males are mostly light gray.

✳ BEHAVIOR
These birds sometimes hover in one spot as they try to zero in on prey.

✳ MORE
The birds have a white rump patch.

SHARP-SHINNED HAWK

 Also called sharpies, sharp-shinned hawks are quick and agile. Like related hawks that prey mostly on birds, sharpies are known as bird hawks.

WHERE TO FIND:
Sharp-shinned hawks live in or near forested regions throughout most of North America.

WHAT TO LOOK FOR:

✱ SIZE
These hawks are 10 to 14 inches long with a wingspan of about 2 feet.

✱ COLOR
Adults have dark, gray-blue backs and orangy breasts.

✱ BEHAVIOR
Sharpies pluck the feathers from their prey before eating them.

✱ MORE
They are among the most numerous hawks in North America.

Short, rounded wings and long, rudder-like tails help sharpies dart through wooded areas after prey.

Young sharp-shinned hawks have heavily streaked and spotted heads and pale eyes.

21

COOPER'S HAWK

 A Cooper's hawk looks so much like a sharp-shinned hawk that even experienced bird-watchers have difficulty telling them apart. Cooper's hawks and sharpies fly with a series of quick flaps and glides.

WHERE TO FIND:
Cooper's hawks live near dense woods from southern Canada through most of Central America.

WHAT TO LOOK FOR:

✴ SIZE
A Cooper's hawk is 14 to 20 inches long; its wingspan may reach 3 feet.

✴ COLOR
The birds have gray-blue backs, orangy breasts, and dark crowns.

✴ BEHAVIOR
They often dart from trees and snatch songbirds with their long legs.

✴ MORE
Sometimes, in speedy pursuit of prey, they crash into large glass windows.

A Cooper's hawk watches for small birds and mammals from a tree branch.

23

NORTHERN GOSHAWK

The northern goshawk (GOS-hawk) lives in North America, Europe, and Asia. It was once called the goosehawk, perhaps for its goose-like call. Goshawk comes from the word "goosehawk."

The northern goshawk is as fierce as it looks. It is one of the most aggressive hunters of the raptor world.

WHERE TO FIND:

The northern goshawk ranges through northern and mountain forests of North America.

WHAT TO LOOK FOR:

✷ SIZE
The bird is 21 to 26 inches long; it has a wingspan of almost 4 feet.

✷ COLOR
It is bluish gray with a black-and-white head pattern and white eyebrows.

✷ BEHAVIOR
Goshawks stubbornly pursue and prey on medium-size mammals and birds.

✷ MORE
They are sometimes called "gray ghosts of the forest."

FIELD NOTES

Northern goshawks often prey on snowshoe hares, which they catch and kill with their sharp talons.

RED-TAILED HAWK

The broad wings, ample tail, and plump body of the redtail mark it and its relatives as soaring hawks. The plumage of redtails varies from light to dark depending on the region they inhabit.

FIELD NOTES

Redtails perch beside roadways, watching for mammals in the grass and others killed by cars.

WHERE TO FIND:
Redtails live in woods, fields, meadows, and prairies from Alaska through Central America.

WHAT TO LOOK FOR:

✳ **SIZE**
Redtails are up to 25 inches long and have a wingspan of more than 4 feet.

✳ **COLOR**
Adults are brown and white with a dark belly band and a rusty red tail.

✳ **BEHAVIOR**
Red-tailed hawks sometimes hang motionless in the air while hunting.

✳ **MORE**
They are one of the most common hawks in North America.

Wings spread wide, red-tailed hawks glide effortlessly on air currents.

27

RED-SHOULDERED HAWK

 The red-shouldered hawk is appropriately named. An adult bird has reddish patches on the upper sides of its wings that distinguish it from all other hawks.

WHERE TO FIND:
The red-shouldered hawk lives in wet woodlands from eastern Canada to Mexico and in California.

WHAT TO LOOK FOR:

＊SIZE
The birds are 19 inches long with a wingspan of about 3½ feet.

＊COLOR
They are chestnut with reddish wing linings and bars on their breasts.

＊BEHAVIOR
Like other hawks, they may use old nests as feeding platforms.

＊MORE
They have black tails with narrow white bands.

FIELD NOTES

Red-shouldered hawks court each other by flying together over their nest and calling loudly.

Brown eyes alert, a red-shouldered hawk watches for prey.

BROAD-WINGED HAWK

 The broad-winged hawk is a chunky, crow-size soaring hawk. When they migrate during the fall, broadwings form huge, circling flocks called kettles.

WHERE TO FIND:
Broadwings nest in wooded thickets of Canada and the eastern U.S. They migrate south for the winter.

WHAT TO LOOK FOR:

* SIZE
Broad-winged hawks are about 16 inches long with a wingspan of about 3 feet.

* COLOR
They are brownish with broad black-and-white tail bands.

* BEHAVIOR
When migrating, they may not eat.

* MORE
Broad-winged hawks eat other birds, lizards, frogs, and insects.

These young broadwings still have bits of down, the soft feathers that help keep baby birds warm.

FIELD NOTES
During migration thousands of broadwings may soar together on rising currents of warm air.

SWAINSON'S HAWK

The Swainson's hawk is the most common soaring hawk of the plains. It has one of the longest migration routes of any hawk. Each year it flies from Canada to South America—then back again.

WHERE TO FIND:
Swainson's hawks breed in Canada and throughout the western U.S., then migrate south.

WHAT TO LOOK FOR:

✳ SIZE
A Swainson's hawk is about 21 inches long with a wingspan of 4½ feet.

✳ COLOR
It has a dark back and cream body with a white throat and a dark breast.

✳ BEHAVIOR
Swainson's hawks form huge flocks during migration.

✳ MORE
They eat meat during the breeding season and feed on insects in the winter.

FIELD NOTES
Swainson's hawks sometimes allow songbirds as neighbors, such as these nesting orioles.

At home in grasslands, a Swainson's eats crickets and grasshoppers as well as mice, gophers, and squirrels.

FERRUGINOUS HAWK

 Its ferruginous (fah-ROO-jin-us)—or rust brown—markings gave this hawk its name. The birds are also called squirrel hawks after their favorite food—ground squirrels.

WHERE TO FIND:
The ferruginous hawk's breeding and wintering ranges lie almost entirely within the western U.S.

WHAT TO LOOK FOR:

✳ SIZE
A ferruginous hawk is about 2 feet long and has a wingspan of more than 4 feet.

✳ COLOR
It is reddish brown on its back and wings and is whitish on its underside.

✳ BEHAVIOR
Ferruginous hawks hunt from perches on the ground, on poles, and in trees.

✳ MORE
Their legs have reddish feathers all the way to the toes.

FIELD NOTES

In warm months, ferruginous hawks often hunt prairie dogs by cruising and gliding low over the ground.

The ferruginous hawk is at home on the range and in other types of dry, open country.

HARRIS' HAWK

Harris' hawk was named for Edward Harris, a friend of the famous naturalist John James Audubon. In the desert, Harris' hawks sometimes nest in tall saguaro (suh-WAH-row) cactuses.

FIELD NOTES

Sometimes two or more Harris' hawks hunt together to catch desert quail and other prey.

The call of the Harris' hawk is a loud, harsh scream.

WHERE TO FIND:
Harris' hawk inhabits desert and dry brushland from the Southwest into Mexico and parts of Central America.

WHAT TO LOOK FOR:

✳ **SIZE**
Harris' hawk is about 21 inches long and has a wingspan of nearly 4 feet.

✳ **COLOR**
It is brown to black with chestnut shoulder patches and white on its tail.

✳ **BEHAVIOR**
Year-old Harris' hawks may help their parents feed new chicks in the nest.

✳ **MORE**
Harris' hawks eat other birds, lizards, mammals, and insects.

OSPREY

 The osprey (OS-pray)—
sometimes called the fish
hawk—eats mostly fish. Ospreys live
throughout the world along rivers,
lakes, bays, and seacoasts. They
migrate south for the winter.

WHERE TO FIND:
Look for ospreys near
bodies of water, where the
birds feed on fish swimming
near the surface.

WHAT TO LOOK FOR:

✳ SIZE
An osprey is about 2 feet long; its
wingspan is 5 to 6 feet.

✳ COLOR
It is mostly brown with white markings
and a dark mask on its white head.

✳ BEHAVIOR
Ospreys build massive nests that may
reach ten feet in height.

✳ MORE
They hover over water to search for fish
and plunge in after prey.

FIELD NOTES

An osprey carries a fish with its head facing forward. This method minimizes wind resistance.

An adult osprey returns to its huge stick nest with a fish dinner for its young.

CRESTED CARACARA

 The crested caracara (CARE-ah-CARE-ah) was named for the sound of its call, which is a harsh cackle. It is the national bird of Mexico.

WHERE TO FIND:
Crested caracaras live in open scrublands from the southern United States through Central America.

WHAT TO LOOK FOR:

✳ SIZE
The crested caracara is about 2 feet long and has a wingspan of about 4 feet.

✳ COLOR
It has a black crest, dark wings and body, and white cheeks and throat.

✳ BEHAVIOR
It spends considerable time on the ground hunting small animals.

✳ MORE
Its long tail has a dark band at the tip.

Crested caracaras feed on carrion as well as on live prey. The birds will attack vultures to steal their meals.

The bare skin on the face of the caracara is usually red-orange. It turns yellow when the bird is disturbed.

41

AMERICAN KESTREL

 The colorful American kestrel (KES-trul) is the smallest falcon—and the smallest day-flying raptor—in North America. Like most falcons, it has long, narrow, pointed wings.

WHERE TO FIND:
Most common North American falcon, this kestrel is found in open country, towns, and cities.

WHAT TO LOOK FOR:

✳ SIZE
The American kestrel is about 10 inches long. It has a wingspan of about 2 feet.

✳ COLOR
It has a rust red back and tail.

✳ BEHAVIOR
It nests in abandoned woodpecker holes, in tree cavities, and in nest boxes.

✳ MORE
In summer it feeds on large insects; in winter it eats small birds and mammals.

The American kestrel hovers over an open area searching for prey, then plunges to the ground.

The American kestrel has two black stripes on each side of its white face.

43

MERLIN

 The merlin is a small, dark falcon. Its swift, powerful flight helps it overtake prey—other birds, bats, dragonflies, crickets, and grasshoppers—and pluck them from the air with its feet and talons.

WHERE TO FIND:
The merlin nests in northern North America. It winters from the southern U.S. through Central America.

WHAT TO LOOK FOR:

✳ SIZE
The merlin is about 12 inches long and has a wingspan of about 2 feet.

✳ COLOR
Adult males are gray-blue above; females and young are dark brown.

✳ BEHAVIOR
Merlins often use the old tree or cliff nests of other birds such as crows.

✳ MORE
Merlins were once called pigeon hawks because they kill pigeons.

Merlins may perch atop a dead tree, a post, or a boulder to watch for prey.

PRAIRIE FALCON

 Fast-flying prairie falcons hunt over vast expanses of treeless terrain. They prey chiefly on other birds, chasing them in the air. These falcons also dive to grab mammals such as ground squirrels and rabbits.

WHERE TO FIND:
The prairie falcon inhabits dry and open country from southwestern Canada into Mexico.

WHAT TO LOOK FOR:

✳ SIZE
The prairie falcon is about 17 inches long; it has a wingspan up to 3½ feet.

✳ COLOR
Adults are light brown above, cream and heavily spotted below.

✳ BEHAVIOR
These falcons will attack eagles, hawks, and owls that fly near their nests.

✳ MORE
The birds have white eyebrows and dark bars on their cheeks.

FIELD NOTES
Prairie falcons lay their eggs directly on ledges in cliffs, bluffs, or canyons. The birds do not build nests.

In young prairie falcons the fleshy area above the bill is blue-gray, as are the legs.

PEREGRINE FALCON

 The peregrine (PAIR-uh-grin) falcon is the swiftest of all falcons. Found on nearly every continent, it nests and roosts on cliffs and bridges and on ledges of tall city buildings.

WHERE TO FIND:

Peregrines range throughout most of North America from the far north through Central America.

WHAT TO LOOK FOR:

* **SIZE**

The peregrine falcon is 16 to 20 inches long; its wingspan is 3 feet or more.

* **COLOR**

Adults have a black crown, gray back, and tan or whitish underparts.

* **BEHAVIOR**

Peregrines use their feet to strike down birds in flight.

* **MORE**

They use their strong talons and hooked bills to kill and tear their prey.

A peregrine
feasts on a
young quail it
has killed.
These falcons
feed almost
exclusively on
birds.

GYRFALCON

The gyrfalcon (JER-fal-ken), largest of all falcons, ranges throughout the Arctic tundra of North America, Europe, and Asia. It flies low, scaring birds from cover and then chasing them down.

WHERE TO FIND:
Gyrfalcons range from the Arctic through Canada. In hard winters they move into northern regions of the U.S.

WHAT TO LOOK FOR:

＊SIZE
Gyrfalcons are about 2 feet long with wingspans of more than 5 feet.

＊COLOR
They are mostly white or gray with grayish brown markings.

＊BEHAVIOR
With blinding speed, gyrfalcons overtake even the fastest waterfowl.

＊MORE
Their tails are longer than those of other falcons.

A white gyrfalcon flies in search of prey. Powerfully built, gyrfalcons have triangular wings that are broad at the base.

BARN OWL

 Barn owls are found on all continents except Antarctica. They roost and nest in barns and other buildings as well as in cavities in trees, caves, and cliffs. Their call is a raspy, often bloodcurdling, hiss.

WHERE TO FIND:
The barn owl inhabits open areas and farmland from the northern United States through Central America.

WHAT TO LOOK FOR:

✳ SIZE
The barn owl is about 16 inches long and has a wingspan of almost 4 feet.

✳ COLOR
Its breast and underparts are whitish; its back and upper wings are orange-yellow.

✳ BEHAVIOR
When perched, it may move its head from side to side to focus objects.

✳ MORE
It has a white, heart-shaped face.

The dish-like shape of a barn owl's face helps direct sounds to the bird's ears.

FIELD NOTES

The barn owl is a farmer's friend. The birds are good at killing rats, mice, and other rodent pests.

SHORT-EARED OWL

 The short-eared owl belongs to a group of owls called eared, or tufted, owls because of the ear-like tufts on their heads. This owl flies with irregular wingbeats, often fluttering like a moth.

WHERE TO FIND:
The short-eared owl breeds from Alaska into the northern U.S. and winters as far south as Mexico.

WHAT TO LOOK FOR:

✳ SIZE
The short-eared owl is about 13 inches long; its wingspan can be more than 3½ feet.

✳ COLOR
Brownish orange to light brown, it has streaks on its breast and a pale belly.

✳ BEHAVIOR
It hunts chiefly at dawn and dusk.

✳ MORE
It lives mostly in marshes, where it searches for rodents in the short grass.

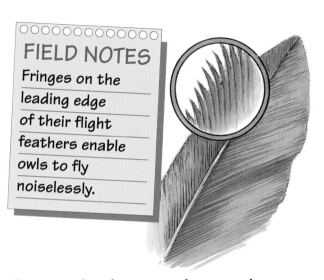

Short-eared owls nest on the ground, usually in shallow depressions that are lined with feathers, straw, and other kinds of material.

LONG-EARED OWL

 The long-eared owl is a slender, secretive owl with long ear tufts that are set closer together than those of other tufted owls. It lives in parts of Europe, Asia, and North Africa as well as in North America.

WHERE TO FIND:
Long-eared owls range throughout southern Canada, the United States, and into northern Mexico.

WHAT TO LOOK FOR:

✳ SIZE
The long-eared owl is about 13 inches long; its wingspan is about 3 feet.

✳ COLOR
It is brownish with bold streaking on its breast and belly, a rusty face, and bright yellow eyes.

✳ BEHAVIOR
It sometimes roosts in large groups in the winter.

✳ MORE
Long-eared owls hunt only at night.

Holding its prey in its talons, a long-eared owl spreads its wings to frighten away enemies.

FIELD NOTES

By raising its ear tufts and stretching its long, thin body, a long-eared owl can hide in plain sight.

GREAT HORNED OWL

 The great horned owl is the largest eared owl in North America. It got its name from its large ear tufts, which look like horns. The call of the great horned owl is a series of loud, deep hoots.

WHERE TO FIND:
These owls live from Alaska through Central America in forests, prairies, deserts, and even cities.

WHAT TO LOOK FOR:

✳ SIZE
The great horned owl is about 22 inches long; its wingspan is almost 5 feet.

✳ COLOR
Forest owls are dark brown; mountain and desert owls are paler and grayer.

✳ BEHAVIOR
A mighty predator, it kills animals as large as skunks and Canada geese.

✳ MORE
It catches prey in its strong feet.

The fierce face of the great horned owl features white eyebrows and a white chin stripe.

59

BARRED OWL

The barred owl is a real hoot—or rather, a real hoot owl. Its call is a rhythmic series of loud hoots: *Who-cooks-for-you, who-cooks-for-you-all*, the last part of which is often dragged out.

Chiefly nocturnal, the barred owl prefers a daytime roost that is well hidden. This owl has large, dark eyes.

WHERE TO FIND:

Barred owls live in woods, often near river bottoms, and in swamps in Canada, the U.S., and Mexico.

WHAT TO LOOK FOR:

✳ SIZE
The barred owl is about 21 inches long; its wingspan is almost 4 feet.

✳ COLOR
It is brown with dark bars on its chin and streaks on its breast.

✳ BEHAVIOR
It takes over old crow and hawk nests and also nests in hollow trees.

✳ MORE
Its diet includes mammals, other birds, and reptiles.

FIELD NOTES

Like other owls, the barred cannot move its eyes. It must turn its head to watch a moving object.

SPOTTED OWL

The spotted owl, the western relative of the barred owl, needs a huge territory in which to hunt for food. It preys chiefly on wood rats and mice. It roosts in thick cover during the day.

FIELD NOTES

In the Pacific Northwest, the spotted owl has lost much of its dense forest habitat to logging.

WHERE TO FIND:

The spotted owl lives in humid, old-growth forests and wooded canyons in western North America.

WHAT TO LOOK FOR:

❋ SIZE

The spotted owl is about 17 inches long; its wingspan is about 3½ feet.

❋ COLOR

It is dark brown with spotted and barred underparts.

❋ BEHAVIOR

Its deep hoot is similar to that of the barred owl.

❋ MORE

The spotted owl hunts only at night.

Like the barred owl, which it closely resembles, the spotted owl has a rounded head and dark eyes.

GREAT GRAY OWL

 The great gray owl looks much larger than it really is. Much of its size is due to its extraordinarily fluffy plumage, which helps keep it warm in its far northern home. Great gray owls often hunt by day.

WHERE TO FIND:

The great gray owl inhabits dense woods and mountain forests in the northern part of the continent.

WHAT TO LOOK FOR:

✳ SIZE
The great gray owl may reach more than 30 inches in length and has a wingspan of up to 5 feet.

✳ COLOR
It is dusky gray with wide streaks on its underparts. It has yellow eyes.

✳ BEHAVIOR
It often shows little fear of people.

✳ MORE
It preys mostly on small rodents.

FIELD NOTES

The great gray owl's unusually large, dish-shaped face helps it hear prey—even under the snow.

Great gray owls often nest in the stubs of broken-off trees.

SNOWY OWL

In their far northern home, snowy owls feed mostly on small rodents called lemmings. In severe winters, when food is hard to find, the owl's prey includes rabbits, hares, squirrels, rats, and other birds.

A snowy owl's white plumage helps camouflage it in its snowy environment.

WHERE TO FIND:
Snowy owls nest on the ground in Arctic tundra. In winter they may range into the northern United States.

WHAT TO LOOK FOR:

✳ SIZE
The snowy owl is 23 to 27 inches long; its wingspan is about 5½ feet.

✳ COLOR
Males are almost pure white; females and young are flecked with brown.

✳ BEHAVIOR
They build nests lined with moss and feathers on raised areas.

✳ MORE
They dive-bomb foxes and even wolves to protect their eggs and young.

FIELD NOTES

Snowy owls have thickly feathered feet that protect them from the cold of their harsh environment.

SCREECH-OWL

Two kinds of screech-owls are common in North America—one in the east and one in the west. Despite its name, the screech-owl doesn't screech. Its call is a soft, mournful whistle or whinny.

FIELD NOTES

In its rufous color phase the screech-owl is North America's only reddish eared owl.

Screech-owls often perch in sunlight at the entrance of a tree cavity.

WHERE TO FIND:

Screech-owls live in forests, wooded areas, towns, and even cities from Canada into Mexico.

WHAT TO LOOK FOR:

✳ **SIZE**
Screech-owls are about 8 inches long and have wingspans of almost 2 feet.

✳ **COLOR**
They occur in different color phases. Gray owls are common in the north and west; red owls in the southeast.

✳ **BEHAVIOR**
Screech-owls like to take baths.

✳ **MORE**
They eat rodents, other birds, and insects.

ELF OWL

 The elf owl is the smallest owl in the world. Despite its sparrow size, it is a fierce hunter. In flight and on the ground, it grabs insects with its feet. It also eats scorpions, removing their stingers first.

WHERE TO FIND:
Elf owls live in deserts, thickets, and wooded canyons in the southwestern U.S. and in Mexico.

WHAT TO LOOK FOR:

✶ SIZE
Elf owls are 5½ to 6 inches long with a wingspan of about 15 inches.

✶ COLOR
They are grayish with yellow eyes.

✶ BEHAVIOR
Males arrive on breeding grounds first and locate a suitable nesting cavity.

✶ MORE
When females arrive to breed and nest, each responds to a male's song and enters his nest cavity.

The tiny elf owl has white eyebrows and soft tan bars and streaks on its breast.

FIELD NOTES

In deserts, elf owls nest in abandoned holes dug by woodpeckers in giant saguaro cactuses.

NORTHERN SAW-WHET OWL

 The northern saw-whet owl was named for its harsh call, which sounds like a saw being sharpened on a whetstone. Active at night, saw-whets roost hidden in evergreen trees during the day.

FIELD NOTES

Their chocolate brown color and plain underparts distinguish young saw-whet owls from adults.

About the size of a person's hand, northern saw-whet owls appear larger in flight, in part because of their fluffy plumage.

WHERE TO FIND:
The northern saw-whet owl lives in or near forests, groves, or thickets from Alaska to Mexico.

WHAT TO LOOK FOR:

✳ **SIZE**
The northern saw-whet owl is only about 7 or 8 inches long; its wingspan is about 1½ feet.

✳ **COLOR**
It is mostly dark brown and chestnut with wide streaks on its underside.

✳ **BEHAVIOR**
It nests in deserted woodpecker holes and tree cavities.

✳ **MORE**
It eats mostly insects.

BURROWING OWL

 The burrowing owl is a long-legged, sociable owl of open country. It often takes over old burrows deserted by rabbits and other mammals. In sandy soil the owls dig their own burrows.

WHERE TO FIND:
The owls live in open areas in western and southern North America and in southern Florida.

WHAT TO LOOK FOR:

✱ **SIZE**
The burrowing owl is about 9 inches long; its wingspan is almost 2 feet.

✱ **COLOR**
Its back is brown with white spots; its underside is white with brown bars.

✱ **BEHAVIOR**
It bobs and bows on its long legs when it is disturbed.

✱ **MORE**
It can often be seen by day standing on the ground or on posts.

Burrowing owls feed mostly on rodents and small reptiles, such as lizards.

Burrowing owls often live in colonies in the abandoned burrows of prairie dog communities.

GLOSSARY

bill The beak or jaws of a bird.

call A sound a bird makes.

carrion Dead and rotting animals.

color phase In birds, the term that describes a different color plumage that may occur in the same kind of bird.

kettle A flock of hawks circling together.

perch A branch or other place where a bird settles, rests, or roosts.

plumage A bird's feathers.

prey An animal that is hunted by other animals for food.

rodent A gnawing mammal, such as a rat, mouse, or squirrel, that has long, chisel-shaped teeth.

roost To rest or sleep, either alone or in groups.

saguaro A tall, tree-like, branching cactus native to Arizona and Mexico.

scavenge To eat dead animals left by other predators or killed by cars.

stoop To dive swiftly after prey.

talons The sharp, curved claws raptors use to catch and hold prey.

territory An area claimed and defended by an animal or a group of animals.

tundra A cold, treeless land on the upper slopes of high mountains and in Arctic regions.

wingspan The measurement of a bird's outstretched wings from tip to tip.

INDEX OF
BIRDS OF PREY

Falcons
American kestrel **42**
crested caracara **40**
gyrfalcon **50**
merlin **44**
peregrine falcon **48**
prairie falcon **46**
Hawks and eagles
bald eagle **14**
broad-winged hawk **30**
Cooper's hawk **22**
ferruginous hawk **34**
golden eagle **12**
Harris' hawk **36**
northern goshawk **24**
northern harrier **18**
red-shouldered hawk **28**
red-tailed hawk **26**
sharp-shinned hawk **20**

snail kite **16**
Swainson's hawk **32**
Osprey **38**
Owls
barn owl **52**
barred owl **60**
burrowing owl **74**
elf owl **70**
great gray owl **64**
great horned owl **58**
long-eared owl **56**
northern saw-whet owl **72**
screech-owl **68**
short-eared owl **54**
snowy owl **66**
spotted owl **62**
Vultures
black vulture **10**
California condor **32**
turkey vulture **8**

ABOUT THE CONSULTANT

George E. Watson was raised in southern New England, where he first became intrigued by birds of prey. He now lives in Washington, D.C. He earned undergraduate, master's, and doctoral degrees from Yale University. Curator of Birds at the Smithsonian Institution's National Museum of Natural History from 1962 to 1985, he is a fellow and past secretary and vice president of the American Ornithologists' Union. He has served on the National Geographic Society's Committee for Research and Exploration since 1975 and is a consultant for Society publications and television productions.

PHOTOGRAPHIC CREDITS